Raquel Martins Porto

Quantitative detection of equine babesiosis by molecular techniques

Raquel Martins Porto

Quantitative detection of equine babesiosis by molecular techniques

Quantification of equine babesiosis by real-time PCR - an alternative to the ELISA test

ScienciaScripts

Cover image: www.ingimage.com

This book is a translation from the original published under ISBN 978-3-330-73747-1.

Publisher:
Sciencia Scripts
is a trademark of
Dodo Books Indian Ocean Ltd. and OmniScriptum S.R.L publishing group

120 High Road, East Finchley, London, N2 9ED, United Kingdom
Str. Armeneasca 28/1, office 1, Chisinau MD-2012, Republic of Moldova, Europe
Printed at: see last page
ISBN: 978-620-7-01288-6

SOROCABA UNIVERSITY
DEAN OF UNDERGRADUATE STUDIES
BIOTECHNOLOGY COURSE

Raquel Martins Porto

QUANTITATIVE DETECTION OF EQUINE BABESIOSIS USING THE QT-PCR TECHNIQUE

Sorocaba/SP

ACKNOWLEDGEMENTS

To all those who collaborated directly or indirectly in the realisation of this work. In particular:

To God, above all, because it is from him that I get the strength and wisdom I need.

To Profª . Ms Renata de Lima, for the opportunity, guidance, teachings and financial support for the development of this work.

To the other Professors of the Biotechnology Course for their support and lessons learnt.

To my classmates for their support and the moments we shared.

To my parents Francisco and Marta for their love, encouragement and support.

To UNISO for its contribution to the study.

To everyone who has been present in my life and during the realisation of this work, thank you very much.

SUMMARY

Equine babesiosis is one of the main parasitic diseases of horses and causes great damage to animal health, so detecting it as soon as possible makes treatment more successful. The aim of this study was to apply the real-time PCR technique (Q-PCR) to quantify equine Babesiosis using primers specific to the *Babesia equi* haemoparasite. The method used was the extraction of blood DNA from horses followed by amplification with specific primers for *Babesia equi* parasites. Two types of test were carried out: nested-PCR and Q-PCR to identify and quantify the parasite. Seven horses were analysed, three of which showed high levels of the parasite, although they had no clinical symptoms of the disease at the time of collection, but fell ill months later. This study opens up the possibility of using the Q-PCR technique to diagnose this equine pathology.

SUMMARY

1 INTRODUCTION

1.1 General Aspects

Babesiosis is caused by intra-erythrocytic infections by protozoa of the genus *Babesia. It* is one of the most common infections of free-living animals and is gaining increasing interest as an emerging zoonosis in humans. Although capable of infecting a wide variety of vertebrates, *Babesia* parasites require both vertebrates and non-vertebrates to maintain the transmission cycle. The parasites replicate in the "red blood cells" of the hosts and when they are in this form they are called **piroplasms** due to their pear-shaped appearance inside the infected cells. (2000, HOMER et al., apud KAKOMA; MEHLHORN, 1994) Equine babesiosis is caused by the protozoa Babesia *caballi* and *Babesia equi, which are* widespread in tropical and subtropical regions. They are transmitted by various species of ticks of the genera *Dermacentor, Rhipicephalus, Hyalomma and Boophilus* (STILLER; COAN, 1995).

The first description of the parasite was made by Guglielmi in 1899, but its determination was made by Laveran in 1901, who named it *Piroplasma equi*. Later, Nuttal and Strickiland in 1910-1912 announced another pathogenic organism, *Babesia caballi,* which was morphologically differentiated from *Babesia equi* and which also causes babesiosis in horses (NANTES; ZAPPA, 2008).

Babesiosis is difficult to control in endemic regions. It is important to prevent the spread of the parasite by using correct diagnosis, effective treatment and tick control (KRATSCHEMER, 2005).

In Latin America, babesiosis is considered an endemic disease. (2004, PEREIRA et al. , apud RISTIC, 1988) In Argentina, between 1970 and 1972, 3 racehorses and jumpers were sold to the United States of America, rejected at the Miami station, quarantined and then returned to their countries of origin at the expense of their owners. This had a major impact on the international horse trade, as the animals were asymptomatic carriers of *Babesia* sp. (2004, PEREIRA et al., apud DURRIEU; CIFOLELLI, 1972)

Ticks transmit a wide variety of pathogenic microorganisms more than any other group of arthropod vectors. Ticks that transmit diseases of protozoan origin can cause (e.g. babesioses and theilerioses) and *ehrlicial* diseases (e.g. anaplasmoses, ehrlichioses) that affect the health of animals, humans and pets worldwide. (JONGEJAN; UILENBERG, 2 004)

According to Holman et al. (1993, apud ENIGK, 1950) equine babesiosis, also known as equine piroplasmosis, is a disease caused by species of haemoprotozoan ticks with a worldwide distribution. The *Babesia* spp. species is able to resist high temperatures within its tick vector, so transmission can occur in both temperate and tropical climates.

Although many PCR assays have been developed to detect a particular pathogen, many pathogens occur together with other species and are transmitted concomitantly, as in the case of the tick vector. These species (which have the same vector) require a universal test where it would be possible to simultaneously and directly detect all the protozoa and parasites likely to be present in the blood of an infected host or in tick vectors (JONGEJAN; UILENBERG, 2004).

2 BIBLIOGRAPHICAL REVIEW

2.1 Parasite cycle

There are around 25 species of babesias recorded in various vertebrate hosts. The aetiological agents *B. caballi* and *B. equi* belong to the Babesiidae family. The Babesia genus can be divided into two groups: the **large species, of** which *B. caballi is a* member, which is more than 3 pm long, and the **small species**, such as *B. equi,* which is 1-2.5 pm long. *B. equi* is considered more pathogenic than *B. caballi.* Babesia equi's definitive host (vertebrates) is horses and its intermediate host is ticks, such as *Amblyomma* spp, *Rhipicephalus sanguineus* and *Anocentor nitens,* among others (KRATSCHMER, 2005). Babesiosis transmitted by the *Babesia equi* species in Brazil is mostly transmitted by the *Boophilus microplus* tick (Figure 1).

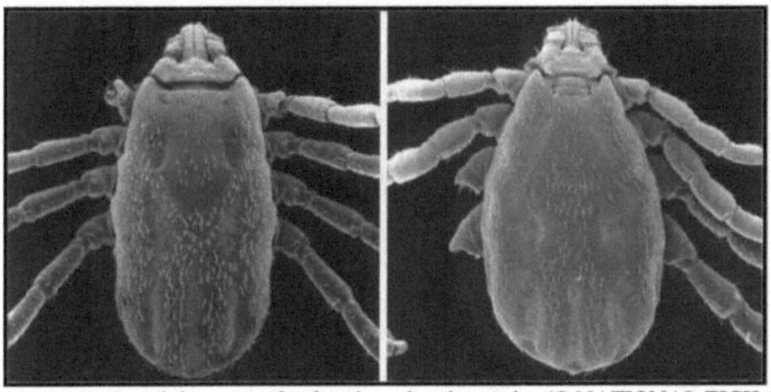

Figure 1 - *Boophilus microplus* female and male species (© NATIONAL TICK COLLECTION, 2004)

Molecular methods have detected a high prevalence of *B. equi* infections in horses, and there is no evidence of diversity between the species of the pathogen isolated in the different regions of Brazil. (HEIM et al., 2007) Transmission occurs through tick bites or contaminated fomites (objects). (SILVA et al., 2005), 2005) The life cycle of the spores of both species has three stages of reproduction: gamogony (formation and fusion of gametes inserted into the tick's intestine), sporogony (asexual reproduction in the salivary glands) and merogony (asexual reproduction in vertebrates). Infection with *B. equi* causes a disease known as nutaliosis or equine piroplasmosis (BRÜNING, 1996).

Infections by *B. caballi* also cause equine babesiosis, but the two agents are quite different and also cause different manifestations of the disease. (NIZOLI, 2005) The life cycle of the babesias differs in terms of infection: *B. caballi* infects erythrocytes, and *B. equi* infects lymphocytes and erythrocytes. During the cycle, the sporozoites are inoculated into the horse's bloodstream by the tick, at which point the cycles of the two babesias are different (MEHLHORN & SCHEIN, 1998).

In *B. caballi,* the inoculated sporozoites turn into trophozoites that grow and divide into two pear-shaped merozoites, which are joined at the back. When a new vector ingests the blood of this host, albeit with low parasitaemia, it becomes infected and thousands of new sporozoites are produced in the acids of the salivary glands of these arthropods (MEHLHORN; SCHEIN, 1998).

B. equi has its cycle with the initial invasion of sporozoites inoculated into lymphocytes, which develop to form macro- and micro-schizonts, which give rise to micromerozoites that are released into the circulation and penetrate the erythrocyte. Micromerozoites are hardly visible under the microscope when parasitaemia is low, and can be mistaken for artefacts because of their size. During the intra-erythrocytic reproductive cycle, they form a tetrad that forms the Maltese Cross, which is more easily seen in acute infections with high parasitaemia. (MEHLHORN; SCHEIN, 1998) Figure 2 shows the characteristic Maltese Cross of *Babesia equi.* (KRATSCHMER, 2005)

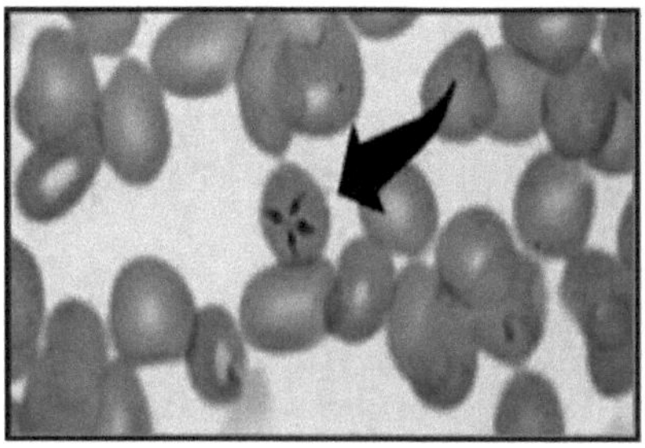

Figure 2 - Maltese cross characteristic of *B. equi* infection (KRATSCHMER, 2005)

There is currently disagreement as to the classification of *B. equi,* as it is often not considered a true babesia due to differences in morphology, ability to infect leucocytes and gene sequence, being considered a *Theileria* and renamed *Theileria equi* (MEHLHORN; SCHEIN, 1998). However, this classification is still uncertain, as *Theileria* does not have an erythrocytic cycle (KRATSCHMER, 2005).

2.2 Clinical Signs

Also in the reproductive area, the disease is recognised as a cause of embryonic resorption and abortions in mares carrying *Babesia* spp. Due to the appearance of fever, sometimes of an intermittent nature, anaemia, jaundice, hepato- and splenomegaly, bulirubinuria and haemoglobinuria may be present in the final phase of the disease (NANTES; ZAPPA, 2008).

Equine babesiosis is a disease that has a major impact on horse breeding, as it is a haemoparasitosis caused mainly by *Babesia equi,* which by infecting and destroying red blood cells causes a drop in animal performance, especially in athletic horses. (ROYES et al., 2005) The prevalence of *B. equi is* independent of the sex and age of the animal. (KRATSCHMER, 2005)

In Brazil, the export of horses would be an important boost for local horse breeding, but unfortunately health barriers are the biggest obstacle to the acceptance of Brazilian animals abroad, because there is no effective control of nutaliosis, piroplasmosis or babesiosis in Brazilian animals (NANTES; ZAPPA, 2008).

Foals are naturally more resistant to infection than adult animals. In endemic areas, young animals have higher antibody titres, indicating a decline in immunity as they get older. The animal can present the disease in an acute, subclinical or chronic form. The clinical signs are: a drop in performance, hyperthermia, anorexia, prostration, anaemia, jaundice, haemorrhagic petechiae on the mucous membranes, oedema of the eyelids and limbs. In extremely serious cases, haemoglobinuria and abortions can occur (KRATSCHMER, 2005).

Most of Brazil is endemically stable, but there are some regions with instability or areas free of *Babesia* spp. Animals born in vector-free areas are highly susceptible to babesiosis. In areas of endemic instability, animals born in climatic periods unfavourable to the vector spend the initial period of their lives without suffering infections, however, they become highly susceptible to *Babesia* infections at advanced ages, when climatic conditions favour the development of the vector and the animals are no longer protected by passive immunity (2006, PEREIRA apud KESSLER et al., 1983).

Many studies show that animals with antibodies against *Babesia* (considered chronic) perform at a lower level than negative animals. The explanation for the decrease in athletic potential is the anaemia that carriers present, even if it is slight, and there can be an acute phase from time to time, which can lead some animals to unexpected drops in performance. Equine babesiosis is one of the main causes of international equine traffic problems, and only 10% of the world's horse population lives in free areas (KRATSCHMER, 2005).

Vertical transmission in ticks occurs in *B. caballi,* but not in *B. equi.* In the acute phase, animals show intravascular and extravascular haemolysis, thrombocytopenia and jaundice, and in the chronic phase they become lifelong carriers, often asymptomatic. Despite the severity of the acute infection, most animals move on from this phase to the chronic phase, which can be re-aggravated by decreases in antibody levels in stressful situations. (KRATSCHMER, 2 005)

2.3 Ways to Detect Babesiosis

The most widely used laboratory technique for detecting infection was microscopic visualisation of the parasites in stained blood smears, but due to their low sensitivity, indirect immunological and enzyme-linked immunosorbent assays (ELISA) have been used (2005, NIZOLI apud CARLLOW et al., 1979).

Haemoprotozoa are routinely tested by light microscopy on blood smears stained with Giensa. However, during the chronic phase of the disease, when the animals generally show non-specific symptoms and parasitemia of less than 0.01%, the sensitivity of this technique becomes very low, resulting in a large number of false negatives. (2005, NIZOLI apud DE WALL et al., 1992)

The ELISA (Enzyme-Linked Immunosorbent Assay) test is used to detect specific antibodies and can be very efficient and sensitive. The quality of the antigens available, however, is crucial for developing the sensitivity and specificity of the test (BOSE et al., 1995). The technique consists of a solid phase capture enzyme immunoassay kit that uses a mixture of synthetic peptides (antigen epitopes) coating a microplate which, in contact with diluted serum or plasma, can form a stable complex to which antibody (IgG) and anti-antibody labelled with peroxidase enzymes are added. It is a generally sensitive and practical technique, but requires several hours to perform and may not be ideal for simple screening due to its high cost, and is directly related to the quality of the antibodies used. (SOARES et al., 2000) The production of antigens for serological reactions has been described from the multiplication of *B. equi* in splenectomised foals with high parasitemia. Native parasite antigens obtained from infected erythrocytes are widely used to detect antibodies in animals with acute and chronic infection (NIZOLI, 2005).

Many countries considered to be free of the disease, such as the United States,

Australia and Japan, aggressively restrict the international transport of animals, whether for temporary or permanent export, and entry requires a negative ELISA test for both *Babesia caballi* and *Babesia equi.* The problem lies in measuring the levels of type G immunoglobulin in the serum of horses. Positivity in these tests means that the animal is considered a carrier, and these immunoglobulins can remain circulating throughout the horse's life, even if contact with new parasites is inhibited. Because of this, there are many questions about the positive results of tests such as complement fixation and cELISA (competitive ELISA), which test for the presence of circulating immunoglobulins (IgM and IgG) in the detection of sick animals. (RONCATI,
2 006)

Since September 2005, the official method for obtaining permission to transport horses internationally to countries free of the disease, according to the OIE (World Organisation for Animal Health), has been the cELISA test, which overlaps with the complement fixation test for detecting *Babesia caballi* and *Babesia equi.* (RONCATI, 2006) Some studies on *Theileria equi* positive animals using the cross-reaction with *Babesia caballi* and *Babesia rodahaine* antigens show that despite its high sensitivity, the test is not very specific, due to the irregularity of the results. (2006, RONCATI apud BOSE; PEIMANN, 1994)

Considering the most commonly used serological techniques, serological differentiation between *B. caballi* and *B. equi* is possible by fixation, and by IFI (indirect immunofluorescence) to a certain extent during the initial phase of infection, but it is not possible by the ELISA test. (KRATSCHMER, 2005)

The implementation of the direct Polymerase Chain Reaction (PCR) technique in the identification and characterisation of Babesia *equi* and *Babesia caballi* infections constitutes a good clinical diagnosis of equine Piroplasmosis (VARGAS et al., 2004). These molecular methods have a high sensitivity and specificity in the detection of *Babesia* infection in peripheral blood and can differentiate species that are not distinguished morphologically (CRIADO-FORNELIO et al., 2003).

2.3.1 Direct parasite detection

Conventional detection techniques such as microscopic and serological techniques are not always used, as they detect a low level of parasitaemia through blood samples, and serological techniques are only detected with antibodies to the specific antigen. DNA markers of *B. equi* and *B. caballi* in blood have been developed, capable of detecting parasitaemia of 0.0028% for *B. equi* and 0.0016% for B. *caballi,* so their use in conjunction with serological techniques can help with detection in animals. Diagnosing babesiosis using microscopy presents several difficulties. Few laboratories are equipped with staff who know parasitology to prepare and evaluate blood smears and the small size of the piroplasm does not allow the organism to be accurately recognised in blood smears (PERSING et al., 1992).

Some genes have been used to investigate the identification of the parasite, this is possible through databases with genetic data of *Babesia* spp. (HOLMAN et al., 2005) With the advent of Molecular Biology, techniques that enable the amplification of DNA *in vitro,* have provided direct diagnosis of various aetiological agents of diseases of sanitary and economic importance for humans and animals (WILLIAMS et al., 1990).

Battsetseg et al. (2002, apud BATTSETSEG et al., 2001) showed the amplification of *B. equi,* and mixed gene fragments from infections of fasting female parasites of *Babesia caballi* and *D. nuttalli,* by *nested* PCR. Due to previous findings showing that 75 per cent of female tick cases in Mongolia are infected with *Babesia* species, fragments of the *B. caballi* gene were amplified, implying that it is the main infection. Several studies have shown *B. equi to* be the most common species, and *B. caballi* is almost always diagnosed together with *B. equi* infection in horses worldwide.

Identifying carrier animals is fundamental for epidemiological diagnosis and establishing a control strategy. PCR in its standard form considerably increases the sensitivity of direct diagnostic tests (OLEGÁRIO et al., 2005).

The isolation of *B. equi* and *B. caballi* by specific and sensitive DNA tests uses a blood sample that detects parasites in carrier animals and may offer a solution to the problems experienced with serological tests, where the horses tested must be free of parasites (KRATSCHMER, 2005).

The breakthrough in PCR technology has been the development of equipment and techniques that allow PCR to be monitored in real time and eliminate the need to detect amplified products by electrophoresis (PEREIRA, 2006).

Molecular methods, similar to the Polymerase Chain Reaction (PCR), present high sensitivity and specificity of peripheral blood spots to detect babesial infection in peripheral blood and can differentiate species that cannot be distinguished morphologically by the spots. The combination of PCR and genetic sequence analysis can increase the information on subspecies or lineages, presenting an advantage in epidemiological studies using molecular methods (PEREIRA, 2006).

Different alternative forms of PCR such as multiplex PCR, nested PCR or seminested

PCR have been developed and applied to diagnostics in order to increase sensitivity and applicability. Multiplex PCR is capable of identifying different species or genera of organisms present in a single sample, using two or more pairs of oligonucleotide primers in a single reaction (FIGUEROA et al., 1992).

Nested PCR is an alternative method that aims to provide greater sensitivity. In this methodology, the amplified product (amplicon) is subjected to a second amplification process using oligonucleotide primers homologous to the internal sequences of the already amplified segment. Generally, the first PCR uses a gene-specific oligonucleotide primer and the second PCR uses a species-specific oligonucleotide primer for the parasite under study. (RAMPERSAD et al., 2003, QUEIROZ et al., 2004) The main previous studies in Brazil show the prevalence of piroplasmosis using recombinant antigens EMA-1 and P48,

were expressed by species of *B. equi* USDA and *B. caballi* USDA originating from the United States (XUAN et al., 2001)

2.3.2 Real-time PCR

Using the PCR technique, it is possible to amplify a segment of DNA exponentially, and after each cycle the amount of DNA in the reaction doubles, duplicating itself with maximum efficiency. The reality is that PCR reactions reach a *plateau* due to the exhaustion of the reagents after many cycles (WITTWER et al., 1997).

Real-time PCR (Q-PCR) is a methodology that allows gene amplification products to be quantified at all stages of a PCR reaction. During Q-PCR, the accumulation of *"amplicons" is* detected in "real time", for each cycle of the reaction, through the excitation of fluorochromes that mark sequence-specific probes or primers used in the reaction (VIEHMANNI et al., 2003).

The product is measured in the exponential phase of the PCR reaction. Traditional methods use post-PCR agarose gels or other detection methods, which are not accurate. Real-time PCR quantifies DNA and RNA easily and is more accurate than past methods (APPLIED BIOSYSTEMS, 2009).

Real-time PCR requires an instrumentation platform containing a thermal cycler coupled to an optical system for fluorescence excitation and emission capture, as well as a computer for acquiring results and final analysis of the reaction. (2007, ALMEIDA; SADDI apud GINZINGER, 2002) Several real-time PCR technologies are available on the market. The two most commonly used systems are *SYBR Green* and TaqMan (BARBANY et al.,

2000).

SYBR Green is a dye that binds to the minor groove of the DNA double strand, making it fluorescent. The advantages of its use include easy handling and low cost. During consecutive PCR cycles, the amount of double-stranded DNA increases exponentially, thus increasing the amount of bound *SYBR Green* and its fluorescence. However, when non-specific amplification or amplification of primer dimers occurs, these are also detected, interfering with the final quantification of the specific reaction products (BARBANY et al., 2000).

A method for nucleic acid detection with real-time PCR uses *SYBR Green* to detect the accumulation of the DNA amplification product. SYBR *Green* provides sensitive detection, but is not specific. The use of SYBR *Green* with instruments that can improve melting curve analysis to determine melting temperature Tm allows the detection of amplification of different products through the % of G and C contained and the increase in amplified product. This is similar, but not equivalent to Agarose gel electrophoresis for separation, as it is mainly based on the increase. Therefore, *SYBR Green assays* are not specific; they can be used in screening assays in species analyses, where the results will be confirmed. (ESPY et al., 2006) *SYBR Green* binds to all double-stranded DNA strands, emitting a fluorescent signal (Figure 3). Detection occurs in the extension step of real-time PCR. The signal intensity increases with the increasing number of product accumulations in the PCR cycle. However, non-specific PCR products and primer dimers also contribute to the fluorescent signal (QUIAGEN, 2009).

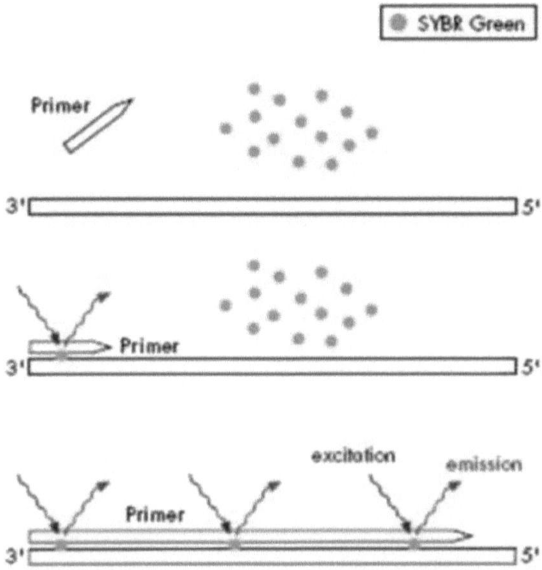

Figure 3. *SYBR Green* principle based on the amplification of products by real-time PCR (QUIAGEN, 2009).

The TaqMan methodology (Perkin Elmer Applied Biosystems) uses the 5'-3' exonuclease activity of Taq® DNA Polymerase and has been widely used. This enzyme digests a labelled probe that specifically anneals to the internal part of the sequence to be amplified between two primers. The TaqMan probe has a fluorescent molecule at the 5' end, which is called a "reporter", and another molecule at the 3' end, which may or may not be fluorescent (of a different wavelength), which is called a *"quencher"*. While the probe is intact, the *quencher* hides the fluorescent signal from the reporter. This effect disappears when the probe, ringed between the primers, is cleaved by the exonuclease activity of the Taq® DNA Polymerase enzyme. While the PCR reaction is taking place, the polymerase synthesises new chains from the primers and cleaves the corresponding probe, resulting in an increase in the fluorescent signal that is picked up with each cycle until a *threshold* is reached at which all the samples can be compared. This threshold corresponds to the moment used to analyse the fluorescence. The threshold is defined in the exponential phase of the PCR reaction, when the amount of product formed satisfactorily reflects the initial concentration of bands amplified by the reaction (HOLLAND et al., 2001).

The most important difference between the *SYBR Green* I system and the TaqMan system is that *SYBR Green* I will detect all double stranded DNA, including non-specific products. An optimised reaction is essential for accurate results (NOVAIS; ALVES, 2004).

The possibility of monitoring PCR in real time has revolutionised the process of quantifying DNA and RNA fragments, as these nucleic acids can be quantified precisely and with greater reproducibility, by determining values during the exponential phase of the reaction (NOVAIS; ALVES, 2004).

The emerging technology of real-time PCR has demonstrated its versatility and usefulness in various fields of biomedical research and biomolecular diagnostics, with the imagination of researchers and manufacturers of new equipment, products and services associated with these branches of knowledge being a limiting factor (DISSAULT; POULIOT, 2006).

Results can be calculated using absolute or relative quantification. The absolute quantification test is used to quantify unknown samples by interpolating their quantities from a standard curve. In absolute quantification, the standards must initially be known by some independent method. A relative quantification assay is used to analyse changes in gene expression in a given sample relative to another reference sample (such as an untreated control sample). (APPLIED BYOSISTEMS, 2009)

In studies on the transplacental determination of *Theileria equi* in Lusitano Thoroughbred equine neonates with 50 foals and their respective dams using the qRT-PCR technique based on the DNA intercalating fluorophore SYBR Green, Rocanti et al. (2006) found that 79.3% of foals were positive and born to mothers who were also positive, demonstrating that the technique is very sensitive and specific, but the possibility of false negatives is considered, and their data demonstrate the possibility of transplacental transmission.

Its use in many laboratories becomes appropriate, simple and realisable through a routine validation of the methodology, avoiding practical limitations (LAURIONOV; KRAUSE; MILLER, 2005).

The data shown in this end-of-course study aimed to detect equine haematozoa by Polymerase Chain Reaction (PCR).

2.3.3 Real-time PCR analysis

Real-time PCR must follow certain guidelines to guarantee reliable results. The statistical analysis used after capturing the fluorescence must be carried out rigorously so that the results show the real situation (Appendix 1). The aim of this study was to detect *Babesia equi* protozoan DNA through the Polymerase Chain Reaction technique, using *nested-PCR,* a variation of PCR, and to quantify *Babesia equi* protozoan DNA through the Real-time Polymerase Chain Reaction (Q-PCR) technique, using parasite-specific primers.

3 OBJECTIVE

3.1 General Objective

The aim of this study was to detect equine haematozoa by Polymerase Chain Reaction (PCR).

3.2 Specific objectives

Detection of *Babesia equi* protozoan DNA through the Polymerase Chain Reaction technique, using ***nested-PCR,*** a variation of PCR.

Quantification of *Babesia equi* protozoan DNA by Real-Time Polymerase Chain Reaction (Q-PCR) using parasite-specific primers.

4 DEVELOPMENT

4.1 Casuistry

The animals under study came from the Sorocaba Horse University. The analyses were carried out on all the animals provided by the university, regardless of the clinical signs presented. Samples from seven animals were used. All stages of this study were approved by the Animal Research Ethics Committee of Univap (Universidade do Vale da Paraíba), protocol no. A26/CEP/2008.

4.2 Method

4.2.1 DNA extraction

The samples were collected in a *vacutainer* with EDTA anticoagulant, where 4 mL of blood were taken from each animal. DNA was extracted from the samples using the protocol of Sambrook et al. (1989) and Barea et al. (2004), for subsequent detection and quantification of babesiosis.

Solution A (Table 1) was added to the blood collected up to a volume of 50 mL and the tube was kept at 4°C for 30 minutes to lyse the red blood cells. The tube was centrifuged at 2,000 rpm, 4°C for 10 minutes. The precipitate ("pellet") was resuspended in 20 mL of **Solution A**. After repeating the operation, the pellet was resuspended in 1 mL of **Solution B** (Table 1), diluted to 1x. 250 pl of **Solution C (Table** 2), containing Proteinase K, was added.

The mixture was incubated under gentle agitation at 37°C for approximately 18 hours or at 65°C for 2 hours. For DNA extraction, 1.25 mL of TE 1x (10 mM Tris-HCl, pH 8.0; 1 mM EDTA) and an equal volume of phenol saturated with 10 mM Tris-HCl, pH 8.0 (APPLIED BIOSYSTEMS, 2009) were added.

The mixture was homogenised by carefully inverting the tube for 5 minutes, after which it was centrifuged at 2,500 rpm for 15 minutes at room temperature to separate and recover the aqueous (upper) phase. An equal volume of phenol saturated with 10 mM Tris-HCl, pH 8.0 was added to the aqueous phase, followed by centrifugation. The procedure was repeated twice with a phenol:chloroform:isoamyl alcohol solution (25:24:1; v:v:v) and finally with chloroform:isoamyl alcohol (24:1; v:v). To precipitate the DNA, 1/10 of the volume obtained was added to 3 M sodium acetate and 2.5x the volume of ice-cold absolute ethanol. The precipitated DNA was recovered using a sterilised plastic rod and washed with 70% ethanol before being resuspended in a volume ranging from 200-500 ui of 1x TE. The DNA concentration was calculated by reading the absorbance at 260 nm on a spectrophotometer, and the integrity of the DNA was tested using 0.8% agarose gel electrophoresis.

Table 1. Solutions used for DNA extraction.

Solution A		Solution B (2x concentrated)	
Triton-X	**1%**	**Na2EDTA**	**20mM**
MgCl2	**5mM**	**NaCl**	**20mM**
Sucrose	**0,32M**	**Tris-HCl**	**20mM (pH 8.0)**
Tris-HCl	**10mM (pH 8.0)**		

The DNA concentration was determined by spectrophotometry at 260 nm using the

Table 2. Proteinase K solution for DNA extraction.

Solution C (for 1mL)
0.5 mL of solution B
1mg Proteinase K (Boehringer Mannheim GmgH, Mannheim, Germany)
0.5 mL of 10% SDS

following information:

- $1A_{260}$ double-stranded DNA unit=50pg

(50pg/ml, with the following DNA concentration calculation)

- DNA (pg/ml)=(OD260)x(dilution factor)x(50pg DNA/ml)/1 unit OD260

Yield= (DNA concentration in pg/ml)x(volume in ml)

4.2.2 PCR

Primers compatible with the *Babesia equi* haemoparasite were developed for PCR analysis. The primers used are described in Table 3, which were chosen according to a study by Aguirre et al. (2004) and Battsetseg et al. (2002).

Table 3. Sequence of the Primers to be used in the work referenced by Battsetseg et al. (2002).

Babesia	**Primer**	**Primer sequence**	**Size**	**References**
	EMA-5(F)	5'TCGACTTCCAGTTGGAGTCC 3'		
B. equi	EMA-6(R)	5'AGCTCGACCCACTTATCACC 5'	268pb	Battsetseg et al., 2001
	EMA-7(F)	5'ATTGACCACGTCACCATCGA 3'		
B. equi	EMA-8(R)	5'GTCCTTCTTGAGAACGAGGT 3'	218pb	Battsetseg et al., 2001

The PCR reaction was developed according to the protocol of Battsetseg et al. (2002) and Aguirre et al. (2004). When determining the specific oligonucleotide primers for *B. equi,* each reaction was carried out independently for each pair of primers. The total reaction was 25pl containing 5pl of DNA sample, 2.5pl of 10X PCR buffer, 2 mM each of dNTP mix, 0.25pl of *Taq* Gold DNA polymerase (5U), 2.5pl of 10pmol of the *B. equi* specific primers, and 12.75pl of sterile ultrapure water. The amplification conditions for the DNA samples were 5 minutes at 94°C and 40 cycles: 94°C in 59 seconds; 58°C in 59 seconds and 72°C in 59 seconds, with a final extension at 72°C in 5 minutes.

For amplification and quantification of the samples in Real Time PCR, the reaction was carried out in 25 pl in which each reaction contained 10.5pL of ultrapure water, 12.5pl of Platinum® *SYBR Green®* (Invitrogen®) and 0.75pl of each pair of oligonucleotide primers. The reaction used 10ng of genomic DNA and a serial dilution of 1:10 was used to determine a standard curve for the primer pairs of *B. equi. /EMA-5*.6 and EMA7.8) in a nested PCR.

We used the conditions already established in the ABI PRISM 7300 Apllied Biosystems machine programme with the SDS 7300/7500/7700 Fast system software, 50°C for 2 minutes; 95°C for 5 minutes; 95°C for 15 seconds, 60°C for 1 minute (40X), adding to this a step of 72°C for 59 seconds and a *hold* at 72°C for 5 minutes, after which we added a cycle to determine the dissociation curve with the conditions of 95°C for 15 seconds, 60°C for 30 seconds and 95°C for 15 seconds.

4.2.2 Visualisation on Agarose Gel

The final amplification product was visualised by electrophoresis on 1% agarose gel in TBE buffer, Blue Green® was used instead of Ethidium Bromide, the image was captured using the Doc. Print. system and Photo-Capt. software to quantify the bands. None of the animals had any symptoms of the disease before the PCR analyses, but all of them were contaminated. The justification for the study was to propose a method for the direct and quantitative detection of the parasite in a sensitive way.

4.2.3 Q-PCR analysis

There are two ways of quantifying samples using real-time PCR: absolute quantification and relative quantification. Absolute quantification is based on the number of copies of the sample of interest related to the signal from the standard curve. Relative quantification is based on changes in the expression of the target gene in relation to a reference group (LIVAK; SCHMITTGEN, 2001).

The fluorescence emitted in response to the synthesis of the amplified product is detected in real time. The increase in this product in the initial cycles is negligible. As the reaction progresses, an increase in the emitted signal can be observed (exponential phase). When this increase in fluorescence becomes statistically significant, the *threshold* level (horizontal line) is reached, and the corresponding PCR reaction cycle is called the *threshold* cycle (Ct). (2008, BARROS; OLIVEIRA; MARIN apud AHMED, 2002) The Ct value can be used to generate a standard curve for a series of samples, where the amount of DNA or RNA is known. (FRACKMAN et al., 2006)

The standard curve is derived from a series of dilutions. Relative concentrations are expressed in arbitrary units as logarithmic concentrations (base 10) that are plotted, which are minimum values used as a standard curve. Calculating the standard curve is a practical method for solving theoretical problems associated with PCR efficiency. The benefit and advantage of the standard curve is being able to assess the limits in data processing and validation of the apparent real-time PCR (LAURIONOV; KRAUSE; MILLER, 2005).

5 RESULTS AND DISCUSSION

5.1 Quantification on agarose gel

In an attempt to quantify babesiosis using simplified methods, nested PCR was carried out followed by a run on a 1% agarose gel.

Agarose gel electrophoresis was used to verify the presence of a single band on the gel corresponding to the amplification of the EMA-5 and EMA-6 primers (Figure 4). This PCR was called PCR-1 and was used as a template for PCR-2 (neted-PCR).

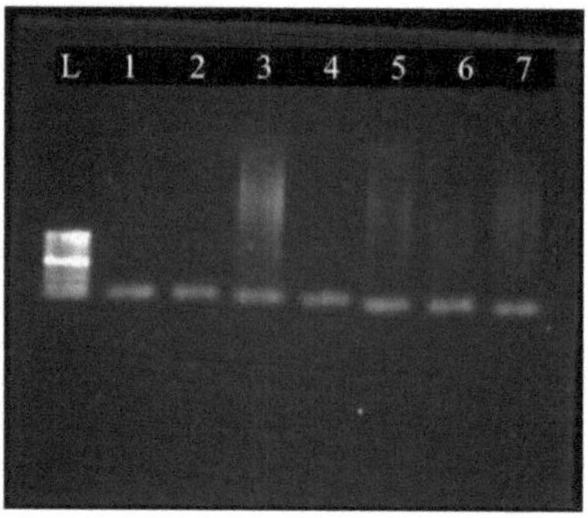

Figure 4 -- Agarose gel showing amplification of *B. equi* DNAs. Selection of the amplified DNA bands referring to the EMA-5 and EMA-6 primers.

Band quantification was carried out using Photo-Capt software (Table 4). The results obtained through the *nested* PCR reaction showed amplification in only four horse DNA samples (Figure 5). The *nested* PCR methodology made it possible to verify the presence or absence of babesiosis, but quantification by Agarose gel did not prove to be an ideal technique for quantification.

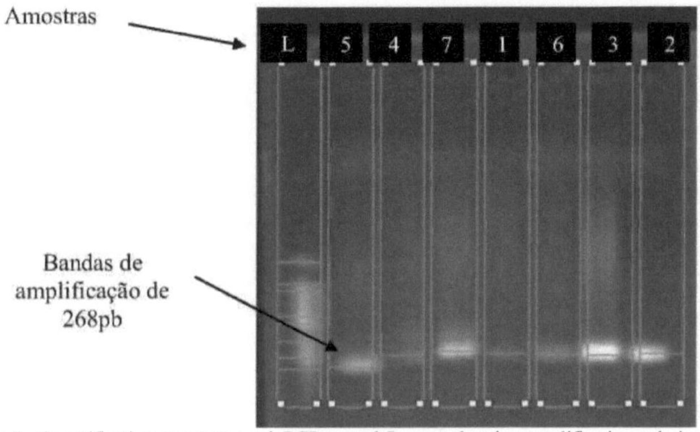

Figure 5 - Quantification on agarose gel. PCR nested *B. equi,* showing amplification only in samples 5, 7, 3 and 2.

Agarose gel quantification showed that horses 3, 2, 7 and 5 were the most contaminated in descending order, with the other horses having practically zero contamination. Horse 1 (least contaminated) was taken as the reference point, with the other values relative to it (Table 4).

According to Table 4, there was a significant difference in the amplification of the EMA-7 and EMA-8 primers between the equine DNA samples evaluated by band volume, brightness and area occupied, but the programme does not show a real quantification of the amount of DNA amplified and is influenced by the conditions of the Agarose gel and its colouring.

Table 4. Relative quantification of the Agarose gel of the horses' DNA samples.

No.	Vol.1 banda	Brilho	Área	imagem
6	2147	19	291	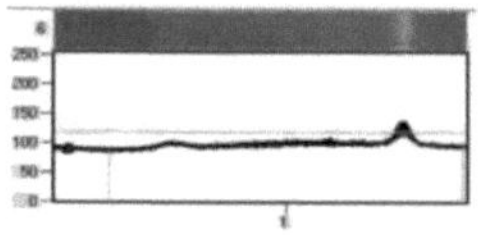
3	45586	135	865	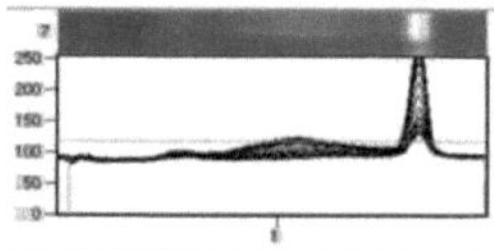
2	22497	124	420	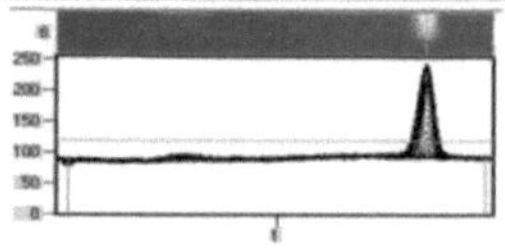
5	11087	55	421	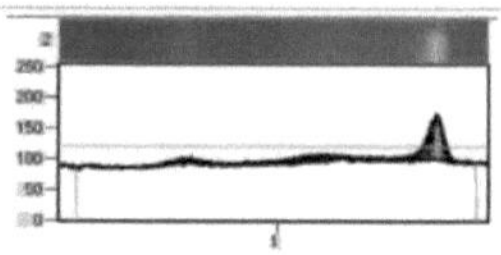
4	480	10	142	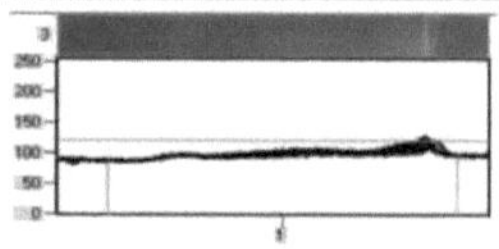
7	24544	91	772	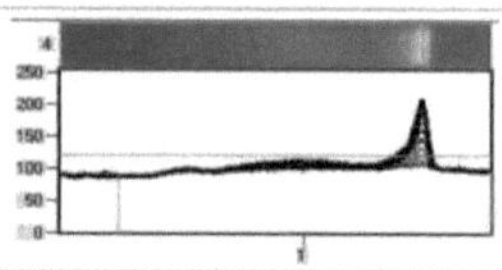
1	1	1	1	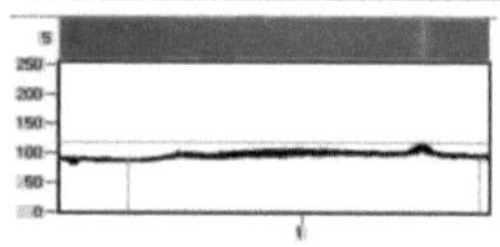

5.2 Real-Time Quantification

5.2.1 *Data reliability*

In order to assess the amplification efficiency, it was necessary to make the experiment's standard curve through successive dilutions. The dilutions used to construct the *standard curve* followed the order: 1/1 (initial sample concentration), 1/10; 1/100 and 1/1000. The efficiencies, calculated by ABI PRISM 7300 Apllied Biosystems with the SDS 7300/7500/7700 Fast system software carried out with the equine 3 DNA sample, can be seen in Figure 6, which represents the efficiency graph, showing a linear relationship between the DNA concentration of the samples used as standards and the Ct values. In the relative quantification method, the amplification efficiencies of the genes of interest and the reference must be similar in order to obtain reliable results (GIULIETT et al., 2001).

Figure 6 - Standard curve of real-time PCR amplification efficiency.

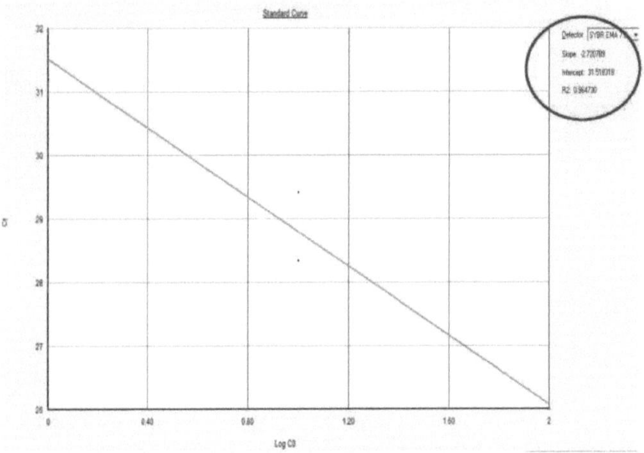

The curve shows values of $R^2 = 0.096$, with efficiency close to 100%. The Slope indicates the inclination of the line and was used to calculate the Efficiency (E), which is given by the formula $E=10^{[-1/slope]}-1*100$.

When calculating the efficiency of the standard curve, a result of 133.099% efficiency was obtained. The value of R^2 is another critical factor for the efficiency of the standard curve. If $R^2 = 1$, the prediction of the value of X (quantity) by the value of Y (Ct) is considered perfect.

by the value of Y (Ct). If R^2 =0, there was no prediction of the value of X (quantity) by the value of Y (Ct). If R^2 >0.99, there is a good correlation between the values of X (quantity) and Y (Ct). The value obtained for R^2 in the test was 0.964730. The recommended efficiency is 90 to 110 per cent. A *slope of* -3.3 ± 10% reflects an efficiency of 100% ± 10%. A PCR reaction with low efficiency will show low sensitivity. (APPLIED BYOSYSTEMS, 2009) These results differ from the efficiency correlation described in the literature and are to be expected due to the lack of a positive contamination sample.

The amplification curves show the TCs for each of the animals, with animal 3 showing the lowest TC value, which indicates more intense amplification, i.e. a greater amount of material (Figure 7).

Figure 7 - Amplification curves of the parasites found in each of the horses. The animals with the highest number of parasites are highlighted.

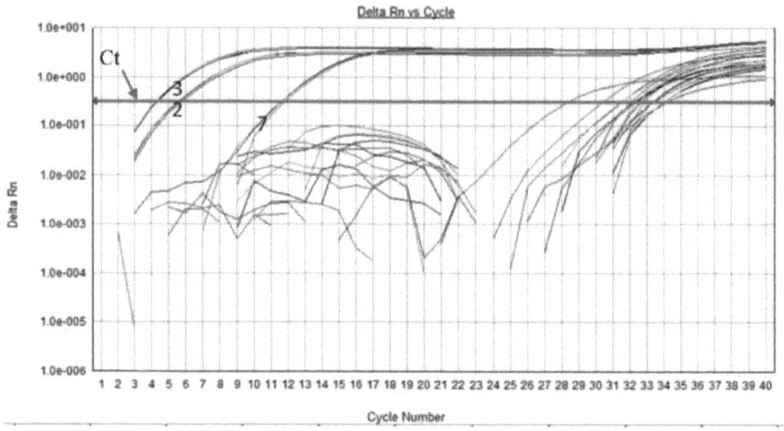

Ct is determined by linearly plotting the logarithmic values of the PCR signal against the number of cycles. Therefore, the CT is the exponential and not the linear value, so one should avoid analysing by Cts values alone. The Ct value is the relative measure of the target concentration in the PCR reaction. Many factors can have an impact on the absolute Ct value in addition to the concentration of the target sample. The statistical analysis used was the calculation of ACt, AACt and 2^{-AACt}. (LIVAK; SCHMITTGEN, 2001)

The comparison between the CTs of each sample was plotted on a graph to make it easier to visualise; the CTs indicate the amplification cycle measured by light emission at the given threshold moment (Figure 8).

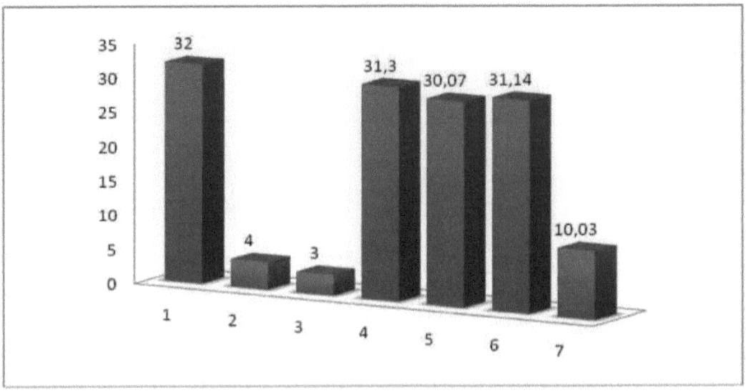

Figure 8 - Figure showing the amplification graph for each sample at a given time. The vertical line determines the CT of each of the samples.

After the amplification reactions were completed, a *"meltin.g"* curve (dissociation curve) was generated for the pair of primers at a temperature of 60°C (Figure 9). This curve allowed the specificity of the amplification to be analysed, where it was verified that the amplified products denatured at the same temperature and that there were no non-specific amplifications or the formation of primer dimers; in some cases, products of different lengths may have the same *"melting"* temperature and are indistinguishable by this analysis. Agarose gel electrophoresis is therefore used to check for the presence of a single band on the gel.

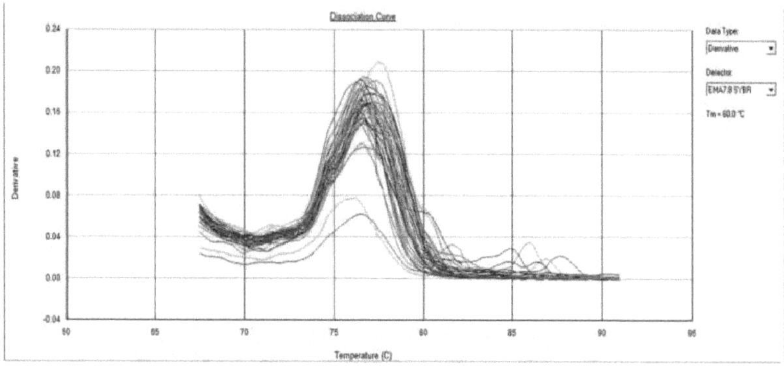

Figure 9 - *Melting* curve of the amplified samples of the EMA7/8 gene at 60°C.

5.2.2 Analysing the results

A sensitive method for assessing the quantification of real-time PCR is to check the ACt (the difference between the two Ct values of two amplifications for the same initial amount of sample) which varies with the dilution of the standard sample. If the efficiencies of the two *"amplicons"* are approximately similar, the plot of the Log of the concentrations versus the ACt will be an approximately horizontal line (slope <0.1). This means that both amplifications have similar efficiencies over the concentration range of the starting sample. If the graph shows unequal efficiencies, the reactions should be optimised until they are similar. (Dussault & Pouliot, 2006) The values found for ACt and AACt result in a relative quantification, which is shown in Table 5. The ACt value of each sample is determined by calculating the difference between the Ct value of the desired gene and the Ct of the endogenous or reference value (QUIAGEN, 2009).

Table 5: ACt and AACt values resulting in relative quantification.

Samples	Average Ct	Deviation	ACt	AACt
1	32	0,33	28,96	25,9
2	4	0,07	0,973	-2,067
3	3	0,02	0	-3,040
4	31,3	0,69	28,26	25,21
5	30,07	0,48	27,03	23,9
6	31,14	0,69	28,1	25,06
7	10,03	0,18	6,99	3,9

The comparative method of 2^{-AACt} was used to compare the relative quantification obtained with the ACt and AACt values, where the quantity of the sample to be quantified is normalised to an endogenous control and relative to a calibrator. (LIVAK; SCHMITTGEN, 2001) Figure 10 shows the graph of the results on a Log 10 basis of the 2^{-AACt} calculation.

Figure 10 - Graph 3. Relative quantification values for the 2^{-AACt} methodology.

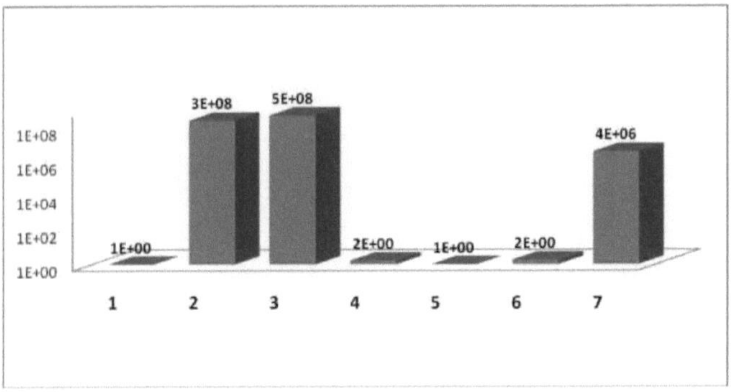

All PCR reaction efficiency methodologies for each sample were derived from standard dilution results. The quantification of the PCR product by fluorescent dye binding depends on homogeneous sample products with specific primers. (APPLEYARD et al., 2006) The results obtained show that horses 2, 3 and 7 have a greater amount of amplified parasilapl DNA, this is verified by the Cts values in which horse 3 would have the greatest amount of parasite in its DNA, since in the third repetition of the cycle it showed fluorescence emission, this result being a relative quantification of the number of copies of the gene. The calculations are based indirectly on comparing the CTs (threshold cycles), which correspond to the PCR cycles in which the exponential amplification signal reaches a fluorescent intensity higher than the detection threshold. The greater the initial amount of genomic DNA, the sooner the accumulated product is detected in the PCR and the lower the CT value (BERNARDI, 2008).

6 CONCLUSION

Babesiosis is one of the main parasitic diseases in horses, causing great economic impact due to the number of animals exposed to infection. The losses associated with *Babesia equi* and *Babesia caballi* infections are related to clinical factors such as preventing the international transit of seropositive animals. Controlling carrier animals is difficult, as after the acute phase there are no signs of infection except the presence of antibodies, making it necessary to detect and visualise the protozoan. Direct diagnosis using the qPCR technique allows the detection and quantification of the protozoan DNA with greater sensitivity, specificity and speed, with the intention of facilitating treatment procedures.

7 BIBLIOGRAPHICAL REFERENCES

AGUIRRE, D.H. et al. "Clinical Babesiosis in horses from Cerrillos, Salta, Argentina". **Revista de Investigaciones Agropecuarias**, v. 33, n. 3, p. 123-133, 2004.

AHMED, F.E. Detection of genetically modified organisms in foods. **Trends in Biotechnology**, v. 20, n. 5, p. 215-223, 2002.

APPLEYARD, Greg D. et al. Differential mitochondrial DNA and gene expression in inherited retinal dysplasia in miniature Schnauzer dogs. **Investigative Ophthalmology & Visual Science**, v. 47, n. 5, p. 1810-1816, 2006.

APPLIED BIOSYSTEMS. **Absolute quantitation using standard curve getting started guide for 7300/7500/7500 Fasts systems**. APPLIED BIOSYSTEMS. Available at: <http:www.appliedbiosystems.com>, accessed on: 18 Feb. 2009.

APPLIED BIOSYSTEMS. **Real Time PCR: Understanding CT**. APPLIED BIOSYSTEMS. Available at: <http://www.appliedbiosystems.com>, accessed on: 18 Feb. 2009.

APPLIED BIOSYSTEMS. **Real Time PCR vs. Traditional PCR**. APLLIED BIOSYSTEMS. Available at: <http://www.appliedbiosystems.com>, accessed on: 18 Feb. 2009.

APPLIED BIOSYSTEMS. **Introduction to real-time PCR**. APPLIED BIOSYSTEMS. Available at: < http:www.appliedbiosystems.com>, accessed on: 27 April 2009.

BARBANY, G. et al. Manifold-assisted reverse transcription-PCR with real-time detection for measurement of the BCR-ABL fusion transcript in chronic myeloid leukaemia. **Clinical Chemistry**, v. 46, n. 7, p. 139-151, 2000.

BAREA, J.A.; PARDINI, M.I.M.C.; GUSHIKEN, T. Extraction of DNA from archival materials and scarce sources for use in polymerase chain reaction (PCR). **Revista Brasileira de Hematologia e Hemoterapia**, v. 26, n. 4, p. 274-281, 2004.

BATTSETSEG, B. et al. Detection of equine Babesia spp. Gene fragments in Demecentor nuttalli Olenev 1929 infesting Mongolian horses, and their amplification in egg and larval progenies. **Parasitology**, v. 1, n. 1, p. 727-730, 2002.

BATTSETSEG, B. et al. Detection of *Babesia caballi* and *Babesia equi* in *Dermacentor nuttalli* adult ticks. **International Journal for Parasitology**, v. 31, n. 1, p. 384-386, 2001.

BERNARDI, C.C. **Study on the amplification and deletion of the BCHE and ACHE genes in sporadic breast cancer patients.** Master's dissertation presented to the Postgraduate Programme of the Genetics Department of the Federal University of Paraná, as a partial requirement for obtaining a Master's degree. Curitiba, 97 p., 2008.

BOSE, R. et al. Current state and future trends in the diagnoses of babesiosis. **Veterinary Parasitology**, v. 57, n.1, p. 61-74, 1995.

BOSE, R.; PEYMANN, B. Diagnosis of Babesia caballi infections in horses by enzime-linked immunoabsorbent assay (ELISA) and Western blot. **International Journal for Parasitology**, v. 24, n. 3, p. 341-346, 1994.

BRÜNING, A. Equine piroplasmosis an update on diagnosis, treatment and prevention. **British**

Veterinary Journal, v. 152, n. 2, p. 139-151, 1996.

CALLOW, L. L et al. Evaluation of an indirect fluorescent antibody test to diagnose *Babesia equi* infection in horses. **Australian Veterinary Journal**, v. 55, n. 1, p. 555-559, 1979.

CRIADO-FORNELIO, A. et al. Characterisation of a *Babesia gibsoni* isolate from a Spanish dog. **Veterinary Parasitology**, v. 1, n. 1, p. 117-129, 2003.

DE WALL, T. Equine piroplasmosis: a review. **British Veterinary Journal**, v. 148, n. 6, 1992.

DISSAULT, A.; POULIOT, M. Rapid and simple comparison of Messenger RNA levels using real-time PCR. **Biology Procediments Online**, v. 8, n. 1, p. 1-10, 2006.

DURRIEU, E.; CIROLELLI, A.V. Piroplasmosis equinas. **Gaceta Veterinária**, v. 34, p. 568581, 1972.

ENIGK, K. The influence of climate on the incidence of piroplasmosis in horses. **Journal Tropenmed Parasitol**, v. 2, n. 1, p. 401-410, 1950.

ESPY, M.J. et al. Real-Time PCR in clinical microbiology: Applications for routine laboratory testing. **Clinical Microbiology Reviews**, v. 1, n. 1, p. 165-256, 2006.

FRACKMAN, S.; KRENKE, B.; HOFFMANN, K. et al. Simple as *SYBR® Green*, Powerful as Probes: The Plexor™ Systems Provide Accurate Quantitation in Multiplex qPCR and qRT- PCR. **Promega Corporation**, v. 1, n. 92, p. 10-13, 2006.

FIGUEROA, J.V. et al. Detection of *Babesia bigemina* infected carriers by polymerase chain reaction amplification. **Journal of Clinical Microbiology**, v. 30, n. 1, p. 2576-2582, 1992.

GINZINGER, D.G. Gene quantification using real-time quantitative PCR: na emerging technology hits the mainstream. **Experimental Hematology**, v. 30, n. 6, p. 503-512, 2002.

GIULIETTI, A. et al. An overview of Real-Time Quantitative PCR: Applications to quantify cytokine gene expression. **Methods**, v. 25, p. 386 - 401, 2001.

HEIM, A. et al. Detection and molecular characterisation of Babesia caballi and Theileria equi isolates from endemic areas of Brazil. **Parasitology Research**, v. 1O2, n. 1, p. 63, 2007.

HOLLAND, P.M. et al. Detection of specific polymerase chain reaction by utilising the 5'-3' exonuclease activity of Thermus aquatius DNA polymerase. **Proceedings of the National Academy of Sciences USA**, v. 88, n. 16, p. 7276-7280, 2001.

HOLMAN, P.J. et al. Comparative infectivity of Babesia divergens and zoonotic *Babesia divergens-like* parasite in cattle. **American Journal of Tropical Medicine and Hygiene**, v. 73, n. 5, p. 865-870, 2005.

JOGEIAN, F.; UILENBERG, G. The global importance of ticks. **Parasitology**, v. 129, n. 1, p. 3-14, 2004.

KAKOMA, I.; MEHLHORN, H. *Babesia* of domestic animals. In: KREIER, J. P. (Ed.). **Parasitic protozoa**. 1994. 2. ed. San Diego: Academic Press, 1994. v. 7, p. 141-216,1994.

KESSLER, R.H. et al. Cerebral babesiosis caused by *Babesia bovis* in calves in the State of Mato Grosso do Sul. **Pesquisa Agropecuária Brasileira**, v. 18, p. 931-935, 1983.

KRASTSCHMER, L. **Equine Babesiosis**. Latu Sensu Specialisation Monograph, University of

Santo Amaro, São Paulo, 2005.

LARIONOV, A.; KRAUSE, A.; MILLER, W. Methodology article: a standard curve based method for relative real time PCR data processing. **BCM Bioinformatics**, p. 1-16, 2005.

LIVAK, K.J.; SCHMITTGEN, T.D. Analysis of relative gene expression data using real-time quantitative PCR and 2-ΔΔct method. **Methods**, v. 25, p. 402 - 408, 2001.

MEHLHORN, H.; SCHEIN, E. Redescription of *Babesia equi* Laveran, 1901 as *Theileria equi* Mehlhorn, Schein 1998. **Parasitology Research**, v. 84, n. 6, p. 467-476, 1998.

NANTES, J.H.; ZAPPA, V. Nutaliosis: Literature review. **Revista Científica eletrônica de Medicina Veterinária**, v. 10, n. 1, p. 679-7353, 2008.

©NATIONALTICKCOLLECTION. ***Boophilus microplus (Canestrini)* Southern Cattle Tick.** NATIONAL TICK COLLECTION. 2004. Available at:<http://www.discoverlife.org/20/q?search=Boophilus+microplus> Accessed on: 29 May 2009.

NIZOLI, L.Q.**"Haematological and humoral alterations in horses exposed to *Babesia equi* infection in the southern region of Rio Grande do Sul"**. Master of Science dissertation. Federal University of Pelotas, Pelotas-Rio Grande do Sul, 2005.

NOVAIS, C.M.; ALVES, M.P. Real-time PCR: a technological innovation in Polymerase Chain Reaction (PCR). **Revista Biotecnologia Ciência & Desenvolvimento.** Ed. 33, 2004.

OLEGÁRIO, A.O.L. et al. Evaluation of semi-nested polymerase chain reaction (SN-PCR) based on the sequence of the *Babesia bigemina* SS-rRNA gene for diagnosis of this haemoprotozoan. **Revista de Patologia Tropical (special supplement)**, v. 34, 2005.

PEREIRA, D.A.A. **Evaluation and optimisation of Polymerase Chain Reactions for molecular diagnosis and epidemiological study of *Babesia bovis*.** 2006. 48p. Saúde Animal-Ciência Animal, Universidade Federal de Mato Grosso do Sul, Campo Grande-Mato Grosso do Sul, Brazil. 2006.

PERSING, D. H. et al. Detection of *Babesia microti* by polymerase chain reaction. **Journal of Clinical Microbiology**, v. 30, n. 8, p. 2097-2103, 1992.

QUEIROZ, P.R.; VALADARES-INGLIS, M.C.; INGLIS, P.W. Survival in soil and detection of cotransformed *Trichoderma harzianum* by nested PCR. **Pesquisa Agropecuária Brasileira**, v. 39, n. 1, p. 403-405, 2004.

QUIAGEN. **Integrated solutions-Real-time PCR applications:** critical factors for successful Real-time PCR. QIAGEN. Available at: <http://www.qiagen.com>, accessed on: 17 April 2009.

RAMPERSAD, J.N. et al. A field evaluation of PCR for the routine detection of *Babesia equi* in horses. **Veterinary Parasitology**, v. 1, n. 1, p. 114: 81-87, 2003.

RISTIC, M (Ed.). **Babesiosis of domestic animals and man. Florida: CRC Press**, p.255, 1988.

ROCANTI, V.N. **Occurrence of congenital *Theileria equi* in Thoroughbred Lusitano foals in Brazil, diagnosed using the RT-PCR technique**. Faculty of Veterinary and Zootechnical Medicine, University of São Paulo. Thesis presented to the Postgraduate Programme in Veterinary Clinic, p. 7, São Paulo, 2006.

ROYES, C.L. et al. Prevalence of *anti-Babesia equi* antibodies in Crioula horses participating in endurance rides. **Revista de Patologia Tropical (special supplement)**, v. 34, 2005.

SAMBROOK J; FRISTSCH, E.F.; MANIATIS, T.E. **Molecular cloning: a laboratory manual.** 2nd edition, Cold Spring Harbor, New York, v. 3, 1989.

SILVA, L.A. et al. Seroprevalence of horses reactive to *Babesia equi*, in the municipalities of Londrina/PR and Pelotas/RS. **Revista de Patologia Tropical (special supplement)**, v.34, 2005.

SOARES, C.O.; SOUZA, J.C.P.; MADRUGA, C.R.; MADUREIRA, R.C.; MASSARD, C.L.; FONSECA, A.H. Seroprevalence of *Babesia bovis* in cattle in the Norte Fluminense mesoregion. **Pesquisa Veterinária Brasileira**, v. 20, p. 75-79, 2000.

STILLER, D.; COAN, E. Recent developments in elucidating tick vector relationships for anaplasmosis and equine piroplasmosis. **Veterinary parasitology**, v. 57 n. 3, p.97-108, 1995.

VARGAS, D.; BONET, R.; OLIVA, P.; CAMPANO, S. Implementation of the PCR technique in the identification of *Babesia* ssp in horses. **Parasitologia Latinoamericana**, v. 59, p. 179-182, 2004.

VIEHMANNI, S.; SCHIEGEL, A.T.; BRUCH, J. et al. Monitoring of minimal residual didease (MRD) by real-time quantitative reverse transcription PCR (RQ-RT-PCR) in childhood acute myeloid leukaemia with AML/ETO rearrangement. **Leukemia**, v. 17, p. 1130-1136, 2003.

XUAN, X. et al. Diagnosis of equine piroplasmosis in Brazil by serodiagnostic methods with recombinants antigens. **The Journal of Veterinary Medical Science**, v. 1, n. 1, p. 1659, 2001.

WILLIAMS, J.G.K. et al. DNA polymorphisms amplified by arbitrary primers are useful as genetic markers. **Nucleic Acids Research**, v. 18, n. 1, p. 6531-6535, 1990.

WITTWER, C.T. et al. Continuous fluorescence monitoring of rapid cycle DNA amplification. **Biotechniques, v.** 22, p: 130-131, 1997.

ANNEX 1

Guidelines for analysing real-time PCR data (QUIAGEN, 2009):

1. Baseline: The baseline is the noise level in the first few cycles, usually measured between cycles 3 and 15, where there is no detectable increase in fluorescence from the amplification of the products, this is non-specific fluorescence reaction.
2. Reporter signal: fluorescent signal generated during real-time PCR, *SYBR Green,* or by a fluorescent probe.
3. Normalised reporter signal (Rn): This is the emission intensity of the reporter dye divided by the emission intensity of the passive reference dye measured in each cycle.
4. Passive reference dye: In some real-time PCR instruments, the dye ROX fluorescent serves as an internal reference for normalising the fluorescent signal.
5. Threshold cycle (Ct): This is the cycle in which the amplification portion crosses the threshold, i.e, in which there is a significant detectable increase in fluorescence.
6. ACt value: Describes the difference between the target Ct value and the Ct gene value corresponding to the endogenous reference gene. Thus: ACt (target gene)-ACt (endogenous gene), used for relative quantification;
7. AACt: Describes the difference between the average ACt value of the sample of interest and the average of the ACt value of a reference sample. The reference sample is also known as the calibrator sample.
8. Efficiency and slope: The slope of a standard curve provides an indication of the efficiency of real-time PCR. A slope of -3.322 means that the PCR has an efficiency of 1 (100%). A slope of less than -3,322 (e.g. -3.8) indicates an efficiency of <1. Generally, most amplification reactions do not reach 100 per cent efficiency due to experimental limitations. A slope greater than -3.322 (e.g. -3.0) indicates PCR efficiency, which appears to be greater than 100 per cent. This can occur when values are measured in the non-linear phase of the reaction or can indicate the presence of reaction inhibitors.

FSC
www.fsc.org
MIX
Papier aus verantwortungsvollen Quellen
Paper from responsible sources
FSC® C105338

Printed by Books on Demand GmbH, Norderstedt / Germany